WHAT IS DENDROLOGY?

Dendrology is the study of trees – a group of plants that have woody trunks, roots, branches and leaves. All trees are plants, but not all plants are trees.

The scientists who study trees are called **DENDROLOGISTS.**

Words that are tricky to understand are in **bold**. Find out what they mean in the glossary.

Words that are difficult to say are in *italics*. Find out how to say them at the back of the book.

CAN TREES TALK TO EACH OTHER?

DISCOVER THE SCIENCE BEHIND DENDROLOGY
(den-DROH-luh-jee)

Written by Olivia Watson
Illustrated by Denis Alonso

Trees are some of the oldest and biggest living things on Earth! They've been around for millions of years, growing across the world and interacting with plants and animals. Towering tall, these gentle giants have lots of secrets.

Dendrologists love to learn the secrets of trees. They look at the clever ways that trees grow, how they change in different **habitats**, and how they help the natural world. So far, they have discovered more than

60,000 SPECIES!

All trees have branches, roots, and a trunk, but – just like people – they can look very different from each other. Dendrologists have lots of questions about trees, but one of the biggest is... how do trees live happily side by side and stay strong?

Baobab tree

Oak tree

Pine tree

Willow tree

Silver birch tree

Cherry blossom tree

One way dendrologists answer that question is by looking underground, where there's a whole secret world. By linking their roots together, trees can share water and **nutrients**, helping out young and ill trees.

Dendrologists were excited to discover that different types of trees team up too, like birch and fir! They get some help from **fungi,** which make links between the trees' roots. This way, the trees can warn each other about dangers such as **drought** or **disease**.

Teamwork happens above the ground too. *Acacia* trees can sense when they're being eaten by animals, and warn nearby acacia trees by releasing a gas into the air...

The other trees then make their leaves taste really bad. This can make some animals ill and keeps the trees safe from **BEING EATEN!**

Some trees even send signals to insects! Dendrologists have found trees that can tell when they're being eaten by a caterpillar. The tree then gives off a smell that attracts wasps, which attack the caterpillar, stopping it in its tracks!

But trees need insects for more than protection. They also need their help to grow fruit, spread their seeds, and make new trees! By growing bright, sweet-smelling flowers, trees show that they have yummy **nectar** ready to eat. This attracts **pollinators** like bees and butterflies that help trees out!

Scent, sight, and **chemicals** might not be the only ways trees send messages! **Botanists** found that some plants make a clicking noise. It could simply be a sound they make when they feel unhappy, like when they're thirsty. But scientists are excited to learn what this discovery means for trees.

Dendrologists have found clever ways of working out how old a tree is, like counting the rings in its trunk. So we now know that some trees live to be thousands of years old! Sending messages from one tree to another is just one of the

SECRETS OF THEIR SURVIVAL.

Trees try to protect each other by sending messages.
But there are still times when they need our help.
Some face **extinction** due to **deforestation**…

some are damaged by forest fires...

and some struggle to fight disease!

Dendrologists show us all the amazing things that trees do for our world. They make homes for animals, clean the air, and keep the soil healthy. Because of all the scientists' hard work, we know it is important to help trees when they can't help themselves.

Thanks to dendrologists, we know that trees talk to other trees and animals too. They just don't talk like us. Instead, they have their very own ways of sending messages.

And that is just one of their many special abilities and features...

Bonsai tree
Living artwork

Great Basin bristlecone pine tree
Oldest tree

Ponderosa pine tree
Fireproof

Cannonball tree
Dangerous fruits

Eucalyptus tree
Rainbow bark

Crooked Forest pine tree
Unique shape

Dragon blood tree
Red **sap**

Hyperion tree
Tallest tree

Árbol Del Tule tree
Widest tree

Tree
SUPERPOWERS

As we have seen, trees are incredibly clever plants that work together to help themselves and each other survive. But that's not all they do! Let's explore their other special features.

Powerful MEDICINE!

Some trees, like alder and willow, can be turned into medicines that heal us when we're ill.

Super SHELTERS!

Lots of animals make their homes in trees, from birds nesting in branches to insects living among leaves.

Fine FURNITURE!

Wood from trees is often used to make furniture like chairs, tables, and beds. How many things in your house are made from trees?

Fresh AIR!

Trees keep our air clean! They take in **carbon dioxide** from the air and release **oxygen**, which humans and animals need to breathe.

Tasty FOOD!

Lots of trees produce yummy fruits and seeds that humans and animals like to eat!

Terrific
TREE FACTS

There's so much to discover about the world of dendrology. Do you know the answers to some of the world's biggest questions about trees?

HOW MANY TREES ARE ON THE EARTH?

There's thought to be more than 3 trillion trees on our planet!

DO DENDROLOGISTS ONLY STUDY TREES?

No! Dendrologists study any woody plants, which includes some shrubs and vines.

HOW LONG HAVE TREES BEEN AROUND?

Fossils show that trees have been on Earth for around 385 million years. This means they were around when dinosaurs lived!

DO TREES GROW EVERYWHERE IN THE WORLD?

Almost! Trees can be found on every **continent** except for Antarctica. They can even grow in extreme places like deserts and swamps.

WHAT HAPPENS WHEN A TREE DIES?

Some trees only live for a few years, but some live for thousands! Even after a tree has died, it's still great for nature, providing food and shelter for animals big and small.

GLOSSARY

Botanists – scientists who study botany (see below).

Botany – the scientific study of plants.

Carbon dioxide – an invisible gas in the air that plants take in to make food and oxygen (see right).

Chemicals – substances made up of the same things, often made by humans.

Continent – one of the huge pieces of land on Earth. For example, Africa and Europe are separate continents.

Deforestation – when forests are chopped down. *Need help saying this? Look below!*

Disease – a condition that causes part of a living thing to no longer work properly. *Need help saying this? Look below!*

Drought – a long period of dryness, usually caused by lack of rainfall.

Extinction – when something like a plant or animal species (see right) no longer exists.

Fossils – the remains or impression of a prehistoric plant or animal.

Fungi – a group of living things, including mushrooms and yeasts, that are neither plants nor animals.

Habitats – the places where animals or plants live.

Nectar – a sugary liquid produced by flowers that attracts pollinators (see right).

Nutrients – substances or ingredients that plants and animals need to live and grow. *Need help saying this? Look below!*

Oxygen – an invisible gas in the air that plants produce, and people and animals need to breathe.

Pollinators – animals and insects that help transfer the small pollen grains needed by plants to make seeds and grow into new plants.

Sap – a watery substance that comes out of a plant or a tree.

Species – a group of living things that share characteristics and features. For example, ash trees and oak trees are different species.

HOW DO I SAY?

Acacia
ah-KAY-sha

Deforestation
dee-fo-ruh-STAY-shun

Dendrologists
den-DROH-luh-jists

Dendrology
den-DROH-luh-jee

Disease
di-ZEEZ

Nutrients
NEW-tree-unts

THE BIG QUESTIONS ANSWERED

This is more than just a series of books; it is a complete resource.
Accompanying each book is a variety of FREE material to engage curious kids with science.

www.thebigquestionsanswered.com

Use the QR code to visit the website, download free resources, and discover other books in the series.

On the website, find out incredible things about dendrologists, including what they do, some of their greatest discoveries, and what it takes to become an expert in this field of science.

The material is also available for home or classroom use, supporting all the information in this book.

Teachers' & Parents' Resources
With discussion prompts and questions, extra information, and facts around key topics.

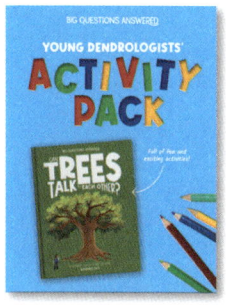

Young Dendrologists' Activity Pack
Fun activities for wannabe tree experts, including creative writing, drawing, word searches, and much, much more.

BEETLE BOOKS

Beetle Books is an imprint of Hungry Tomato Ltd.

First published in 2024 by Hungry Tomato Ltd
F15, Old Bakery Studios, Blewetts Wharf, Malpas Road,
Truro, Cornwall, TR1 1QH, UK.

ISBN 9781835691281

Copyright © 2024 Hungry Tomato Ltd

With thanks to:
Editor: Holly Thornton
Editor: Millie Burdett
Senior Designer: Amy Harvey
Tim Cook for his valued contribution
The team at Beehive Illustration

Printed and bound in China.

Picture Credits:
(t = top, b = bottom, m = middle, l = left, r = right)
Shutterstock: Anitasstudio 32mr; COLOMOBO NICOLA 34bl; Monticello 33br; Creative Travel Projects 34m; Karel Bock 35bl; Kostiantyn Ivanyshen 35tl; Perfect Lazybones 35mr; Studiovin 33ml.